HEALTHY KIDS

UNDERSTAND PUBERTY

by LEON GRAY

Illustrated by DAVIDE ORTU

W
FRANKLIN WATTS
LONDON • SYDNEY

First published in Great Britain in 2024 by Hodder & Stoughton

Copyright © Hodder & Stoughton Limited, 2024
(Text has previously appeared in *Being Healthy, Feeling Great: Puberty* (2009)
and has been updated for this edition.)

Credits
Series Editor: Amy Pimperton
Series Designer: Peter Scoulding
Consultant: Dr Kristina Routh, Sue Beck MSc, BSc

ISBN: 978 1 4451 8864 5 (hardback)
ISBN: 978 1 4451 8865 2 (paperback)

Printed in Dubai

MIX
Paper | Supporting
responsible forestry
FSC® C104740
FSC
www.fsc.org

Franklin Watts
An imprint of
Hachette Children's Group
Part of Hodder & Stoughton
Carmelite House
50 Victoria Embankment
London EC4Y 0DZ

An Hachette UK Company
www.hachette.co.uk
www.hachettechildrens.co.uk

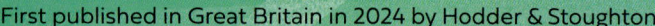

CONTENTS

NOTE:

If you have any worries around puberty, changes to your body, hygiene and your mental health, please speak to a trusted adult, such as a parent, carer or teacher. They can help you to seek advice from a doctor or other medical professional.

WHAT IS PUBERTY?

Puberty is the time when your body changes from being a child to an adult who may be able to choose to have children of their own. Everyone goes through puberty; it is a natural and healthy part of growing up.

PUBERTY FACT

During puberty, you may grow very quickly. You may even suddenly overtake your parents or carers in height!

WHEN WILL PUBERTY START?

On average, puberty starts between the ages of eight and thirteen for girls, and ten and fifteen for boys. Puberty doesn't happen overnight because it is a gradual process. Some people go through puberty faster than others, but it isn't important how quickly or slowly you go through it, or how early or late it begins. Your body will start to change exactly when it is ready and at a rate that is right for you.

A ROLLER COASTER RIDE!

During puberty, your body will grow and change shape and you will experience new thoughts and emotions. Your mood can change from one extreme to the other (see page 7). It can feel like a bit of a roller coaster ride!

DON'T WORRY ...

It's natural to worry about all the changes that happen during puberty. You may worry about the physical changes puberty brings or you may feel confused about the strong feelings you have. If you are worried about anything at all, it helps to talk to someone you can trust, such as a parent, carer, teacher or other trusted adult. They can help you work through any concerns.

CHANGES

HEALTHY HORMONES

Your body is full of natural, healthy chemicals, called hormones. Throughout your life, hormones help your body to work as it should. Growth hormones tell your body to get bigger. Sleep hormones help your body to rest. Hormones play such a big part in your body that you could not live without them.

PUBERTY HORMONES

Hormones are responsible for helping you through the process of puberty. Special puberty hormones tell the different parts of your body to grow and change shape.

They cause lots of other physical changes, too. Without puberty hormones, you would never become an adult. Compare a photograph of yourself now with one taken when you were younger. How have you changed?

THOUGHTS AND FEELINGS

Puberty hormones not only change your body, but they also change the way you think and the way you feel about yourself and others.

Sometimes, during puberty, hormones may cause mood swings, which might make you feel happy one minute and upset the next. Try not to worry – these emotional ups and downs are natural and should settle down over time.

BOYS AND GIRLS

Different hormones work on different parts of the body during puberty. The main puberty hormones are oestrogen, progesterone and testosterone. Boys and girls have all of these hormones in their bodies – but in different amounts. Girls have more oestrogen and progesterone and boys have more testosterone. This is what makes girls and boys look and act differently when they go through puberty.

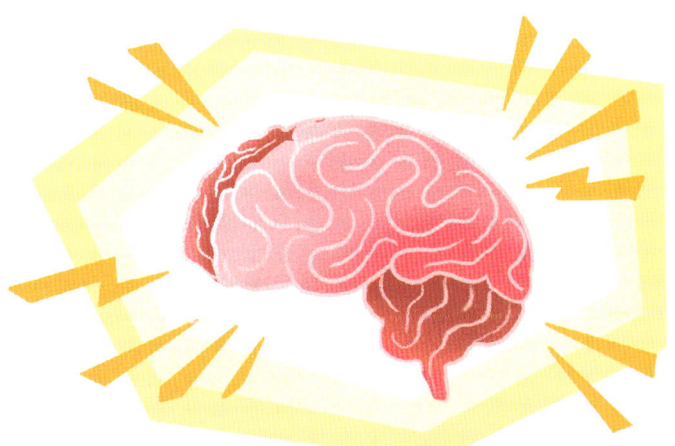

PUBERTY FACT

The brain is the body's control centre. It tells parts of your body – called glands – to make different hormones. The glands then release the hormones into your blood. The hormones can then act on different parts of your body.

GROWTH SPURTS

Something that is usually easy to see during puberty is a rapid change in your height. This is called a growth spurt, and is triggered by your hormones causing you to grow at a much quicker rate than before.

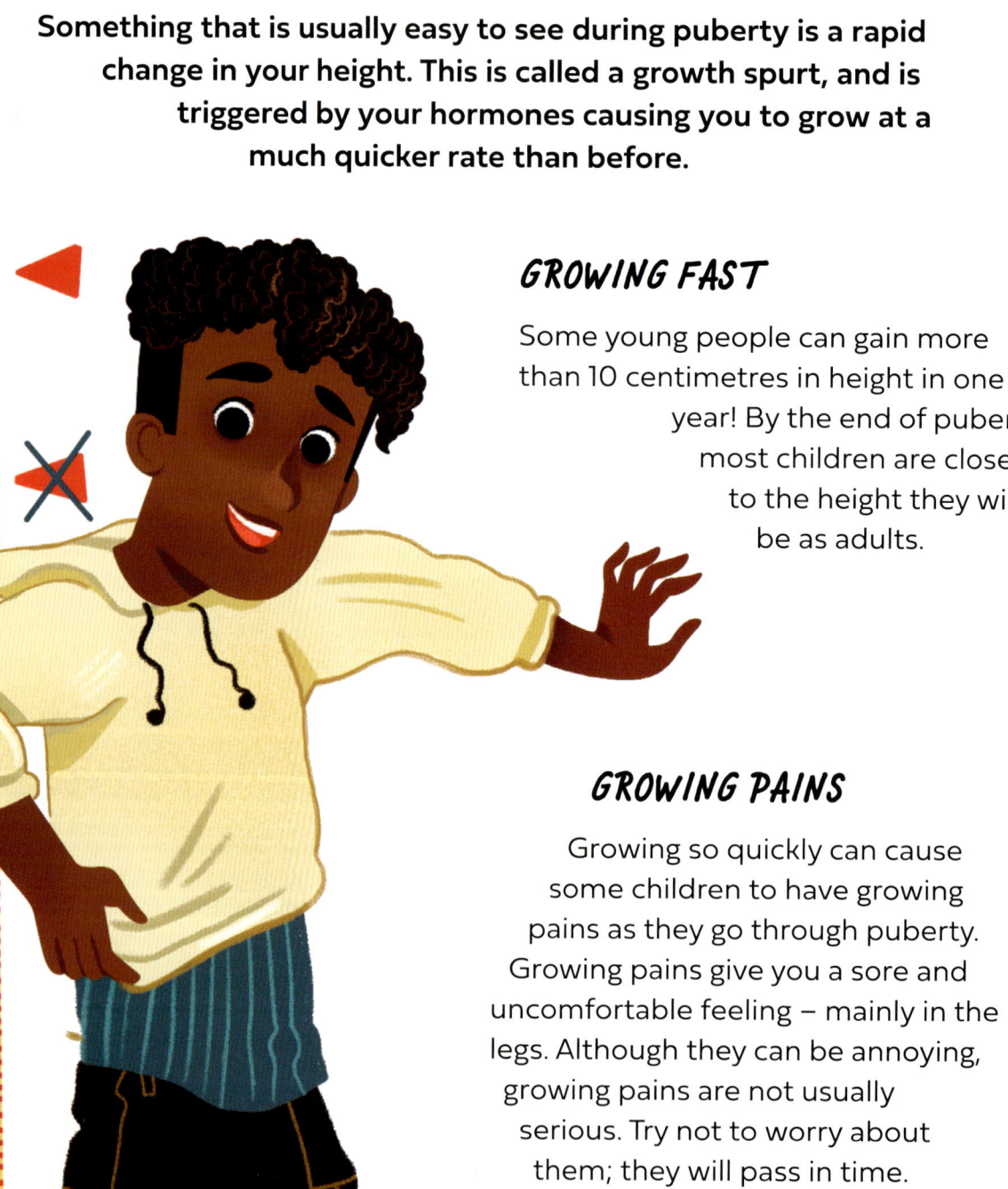

GROWING FAST

Some young people can gain more than 10 centimetres in height in one year! By the end of puberty, most children are close to the height they will be as adults.

GROWING PAINS

Growing so quickly can cause some children to have growing pains as they go through puberty. Growing pains give you a sore and uncomfortable feeling – mainly in the legs. Although they can be annoying, growing pains are not usually serious. Try not to worry about them; they will pass in time.

PAIN RELIEF

If you have growing pains, you can help to relieve the pain by stretching and rubbing the sore area, or using a heat pad. Resting can also help you to feel a little better.

CHANGING SHAPE

Hormones not only make people get taller during puberty, but they also cause their bodies to change shape. Both boys and girls gain weight as they grow taller. Boys' muscles get bigger, and their shoulders become broader. Girls' hips widen and they start to grow breasts.

Changes in your shape, size and weight during puberty are perfectly normal and healthy. Do not feel tempted to cut down on food or to do too much exercise to stop growing bigger. This can be very bad for you.

HEALTHY HINT

Feeling self-conscious?

During puberty, many young people feel self-conscious at times.
If people make unwelcome comments about changes to your body, politely ask them to stop.

BODY HAIR

Puberty hormones also cause changes in the amount and thickness of your hair. During puberty, hair starts to grow on some parts of your body where there was no hair before.

HAIR EVERYWHERE

Nearly everyone has hair on their arms and legs. When you are a child, this hair is often so fine that you can hardly see it. When you go through puberty, it gets thicker and may get darker. New hair grows under your arms, too. At first, this is thin and light, but it gets thicker as you get older.

PUBIC HAIR

During puberty, more hair also grows around your genitals. This is called pubic hair. When it first grows, it is quite thin. Later, it gets thicker and curlier.

HEALTHY HINT

To shave or not?

Many people choose to shave the darker and thicker hair on their legs and under their arms. But you don't have to – the choice is up to you!

FACIAL HAIR

You may have noticed that men often have hair in places that women usually don't. During puberty, boys grow hair on their face, chest and back. This is because boys have more testosterone than girls.

Facial hair often starts with a thin moustache on the upper lip. Later, this hair gets thicker, and hair also starts to grow on the cheeks, chin and neck. Some boys choose to shave their facial hair, unless they want to grow a beard.

EVERYONE IS DIFFERENT

Body hair is nothing to be embarrassed about. Everyone has it. Some people grow more hair than others – everybody is different. People all over the world have different colours, textures and amounts of hair.

HEALTHY HINT

Take care

Take care shaving with razors as the blades are very sharp. A trusted adult must buy a razor for you, because children under the age of 18 are not allowed to buy them. Ask your parent or carer to show you how to shave with a razor, safely.

SPOTS AND ACNE

During puberty, different hormones cause the skin to produce an oily substance called sebum. If the skin produces too much sebum, it can make hair and skin greasy. Washing your hair helps to wash the greasiness away. Sebum can also cause spots and sometimes acne throughout your teenage years.

SPOTTY SKIN

Most people get some spots, especially in the early years of puberty. They often appear on the face, the upper chest and the top of the back. This is perfectly normal and nothing to worry about. You may notice that many of your friends have spots, too.

ACNE

When a person gets lots of spots and they keep coming back, this is called acne. It is very common during puberty. Acne may be caused by a build-up of sebum, blocked pores or a bacterial infection. Acne can make the skin feel very sore and uncomfortable. Fortunately, there is a lot you can do to treat the condition.

WHAT TO DO

The best way to look after your skin if you have spots or acne is to wash your face twice a day using just a mild soap and warm (not hot) water. Do not try to squeeze spots, especially if they are red and sore. This will make them worse.

If you wear make-up, ensure it is completely removed before you go to bed. It's also important to eat healthily and drink plenty of water.

If you are worried about your acne or if it is severe, ask a parent or carer to make an appointment with a doctor.

HEALTHY HINT

Sebum

You don't need to wash away all of your skin's sebum! The body needs some sebum to keep the skin soft and to protect your hair.

BREASTS AND BRAS

One big change for girls during puberty is that their breasts develop. This is another step on the way to your body changing from a child's body into an adult's body.

STARTING TO GROW

When the breasts first start to grow, the nipples may get darker and stick out more. A small swelling will start to grow under them. This might feel a little tender to begin with, but it will soon settle down. Sometimes, the swelling may appear under only one nipple at first.

FAST AND SLOW

As the breasts continue to develop, the whole area around the nipple gets bigger and rounder. Sometimes, one breast may grow a little faster than the other. This is perfectly normal, and the breasts will eventually even out to be about the same size. Remember, it is not at all unusual to have breasts of slightly different sizes.

FEELING SELF-CONSCIOUS

Some girls' breasts start to grow before their friends' breasts. This can make them feel self-conscious or embarrassed. Sadly, some girls may be teased at this time. Ignore any teasing. If teasing becomes overwhelming or more like bullying, always tell a trusted adult.

Remember, everyone changes at different rates during puberty. All girls will develop breasts, because this is a natural part of growing into a woman. But this will only happen when the body is ready – and not before.

SHAPES AND SIZES

Breasts come in lots of shapes and sizes. It is not at all important whether you have large or small breasts. Whatever their size, when you are an adult, you should still be able to feed your baby breast milk, if you choose to.

PERIODS

An important part of puberty for girls is periods. Some girls start their periods as early as nine; others may not start until they are around fifteen. Remember, your periods will start when your body is ready.

WHAT ARE PERIODS FOR?

Women's bodies have eggs stored in their ovaries, which can develop into babies when they are fertilised by sperm from men's bodies (see pages 18–19). Once a month, an ovary releases an egg and the uterus lining thickens. If the egg gets fertilised, it travels to the uterus, where it can develop into a baby. If the egg is not fertilised, the uterus lining is shed and flows out of the vagina. This is a period.

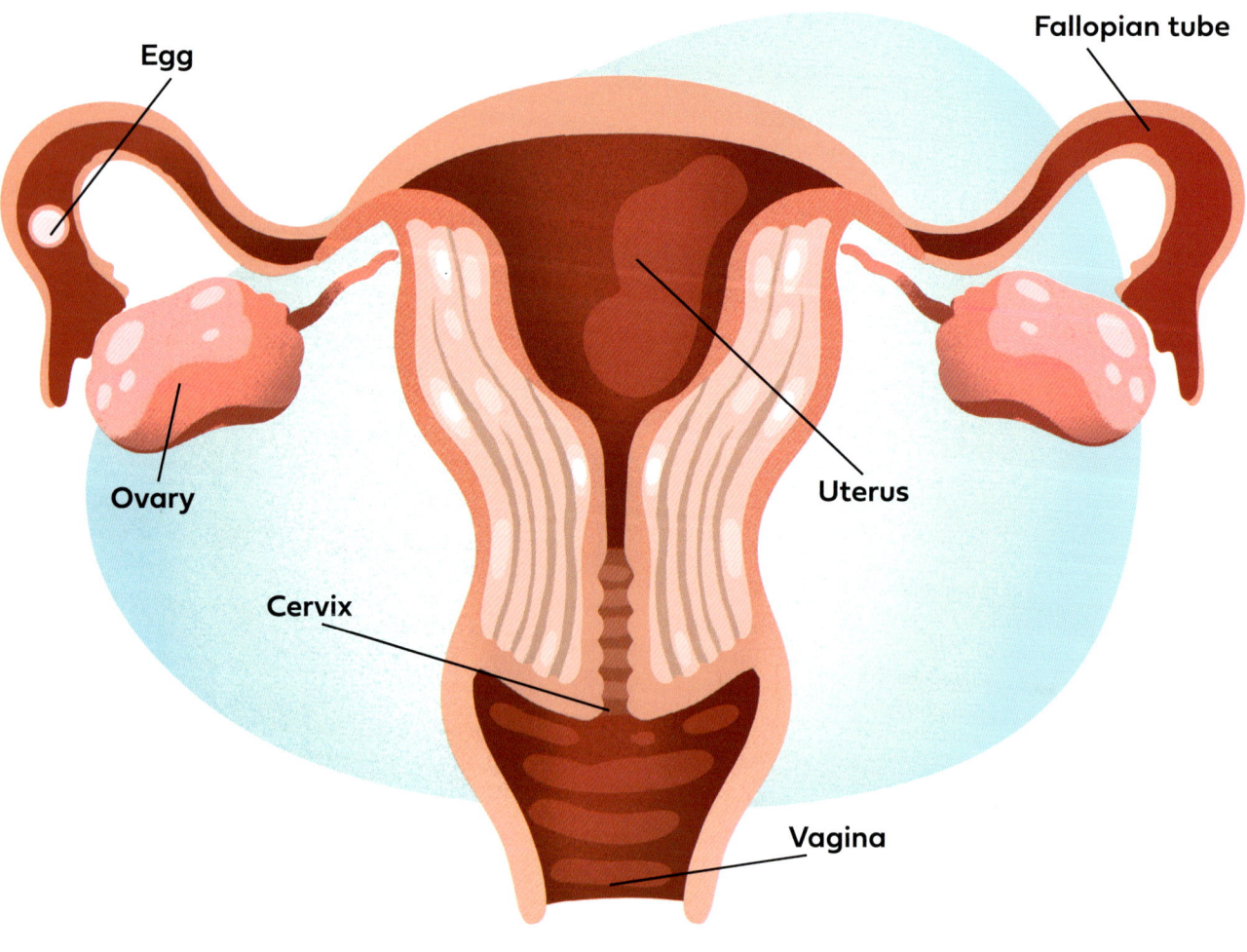

Egg

Fallopian tube

Ovary

Uterus

Cervix

Vagina

SANITARY PRODUCTS

A period usually lasts between three and seven days. During this time, many women and girls choose to use sanitary products, such as pads, tampons or special underwear to absorb the blood from the uterus lining. You can also use silicone menstrual cups that collect your period. You then simply empty the blood into a toilet or sink, and wash the blood away.

PERIOD PAINS

Sometimes, periods feel uncomfortable. You may have a headache, backache or an ache in your lower abdomen. You may feel more tired than usual, or that your emotions are up and down. These feelings should pass.

You can make yourself feel more comfortable by putting a heat pad on your back or abdomen. If you feel upset, talk to a parent or to another trusted adult.

HEALTHY HINT

Change often

Change sanitary products regularly – around every two to six hours. This will help you to feel clean and comfortable.

Never leave a tampon or menstrual cup in for more than eight hours as this can lead to a dangerous condition called toxic shock syndrome. It is rare, but must be treated quickly to prevent harm.

BOY STUFF

During puberty, hormones make a boy's penis and testicles grow. The penis gets longer, and the testicles get bigger and start to produce sperm. Sperm can fertilise an egg from a woman's ovaries to make a baby (see pages 16–17).

SIZE DOESN'T MATTER

Sometimes, boys get embarrassed about their penis size. They may think it is too small or too big, or worry that they may be teased about it. The size of your penis isn't something to worry about. Whether a penis is big or small, it will still allow a man to have children when he is ready.

WET DREAMS

An erection is when blood flows into the penis, causing it to enlarge and harden. Sometimes, during puberty, boys get an erection when they do not want to. Sperm can leak out of your penis when you are asleep. This is called a 'wet dream'. Erections and wet dreams are perfectly natural parts of growing up. Try not to feel worried or embarrassed. These things are simply part of the process of becoming a man.

VOICE CHANGES

Another change that takes place in boys during puberty is that the voice gets deeper. This happens because hormones help to make the voice box bigger. As it grows, you may see it start to stick out at the front of your throat. People often call this an 'Adam's apple'. A girl's voice box is much smaller, and doesn't usually stick out in the same way.

PUBERTY FACT

During puberty, a boy's voice can sometimes sound a bit squeaky. Try not to worry – your voice will sort itself out in time as your voice box adjusts to its bigger size.

CONFUSING TIMES

Puberty hormones can create some strong emotions (see pages 6–7). One minute, you might feel happy. The next minute, you might feel sad or angry. Many young people find it difficult to understand and cope with these emotional changes.

YOU'RE NOT ALONE

If you are finding puberty a difficult time, try not to worry, because you are not the only one. Most people going through puberty will feel the same way at some stage.

UNIQUE YOU

Some of the physical changes that happen during puberty may make you feel confused or self-conscious about your body. These feelings can get worse if you spend too much time comparing yourself to other people, especially people you see on social media.

Relax and be yourself. There is no one 'right' way to go through puberty. Everyone's journey into adulthood is as unique as they are.

NEW RELATIONSHIPS

People often make new friends as they go through puberty and grow older. Sometimes, two people like each other so much that they become much closer. This is a new kind of relationship called a romantic relationship. Being close to one special person also prepares young people for the loving relationships they will have as adults.

It might seem strange if your best friend suddenly starts spending lots of time with someone else. You might feel hurt and left out. Try to see it from both sides, if you can. Your friend is just enjoying the exciting experience of growing up. And, one day, you will likely also have a romantic relationship with someone you like.

HEALTHY HINT

No pressure

Never feel pressured to 'go out' with someone if you don't want to or just because everyone else is doing it. You have your whole life to form relationships – there's no rush!

FAMILY MATTERS

Being independent is a natural part of growing up. It is how people prepare to look after themselves as adults. During puberty, you may feel a strong desire to become more independent. Sometimes, this can lead to arguments with family members.

INDEPENDENCE

Family is an important part of life and families can often be a great source of love, support and fun. Young children learn a lot by following instructions from their parents or carers.

As you grow older, you develop more of your own thoughts and opinions. During puberty, hormones make these feelings much stronger. It is normal to want more freedom, independence and personal space at this time.

CONFUSED FEELINGS

During puberty, there are times when you feel quite grown-up, but also times when you still feel like a child. Sometimes, you might want to be left alone but, the next minute, you might want to be with others. This can feel very confusing. It can be confusing for family members, too!

KEEP TALKING

Communication is the key to happy relationships, not just in the family, but with your friends, too. This means listening as well as talking. It involves respecting other people's views, as well as your own. Always talk to your family and friends about how you feel. They all have different experiences of life – remember, they went through puberty, too! They can help you to make good choices about your own life.

Remember that your family should want only the best for you. It can be hard, sometimes, for parents and carers to accept that you are growing up. But you can work things out if you keep talking!

HEALTHY HINT

Sibling solidarity!

An older brother, sister or cousin who went through puberty recently could be a great person to support you through puberty.

HEALTHY EATING

Eating healthy food is important during puberty. A good diet will help your body and mind cope with all the growing and changing.

BEST FOR YOUR BODY

Your changing body needs you to eat food regularly. A healthy breakfast will give you energy to start your day well. Eat meals at regular times and when you feel hungry. Stop eating when you feel full. If you feel tired or low, improving your diet may help.

A VARIED DIET

Try to always eat a variety of healthy foods. Your growing body needs plenty of carbohydrates, such as cereal, bread and pasta, for energy. Fruit and vegetables give you lots of vitamins. Meat or vegetarian alternatives, such as beans, give you protein, which is great for your muscles. Dairy products, such as cheese, milk and yoghurt, help to keep your bones strong and healthy.

FOODS TO AVOID

Fatty or sugary foods and drinks don't contain many nutrients, so are best saved for occasional treats, rather than eaten every day.

SHAPE AND SIZE

Everyone gains some weight during puberty. Your body needs to change its shape and size for you to become a healthy adult. Gaining weight can make some people feel anxious. Try to remember that it is important to be a healthy weight for your height as you grow up.

EATING PROBLEMS

Some people may have serious eating problems, such as eating too little (anorexia) or binge eating (bulimia). These conditions can be dangerous. Eating disorders can affect both boys and girls.

HEALTHY HINT

Speak up

If you think that you may have an eating disorder or if you are worried about your weight, speak to a parent, carer or other trusted adult, such as a teacher.

KEEP MOVING

Taking exercise is very important during puberty. It helps you to be fit and healthy. Exercise builds strong muscles and bones and it can be a great boost for your mental health, too!

KEEPING IN SHAPE

Puberty is a great time to work on your fitness and form good lifelong health and exercise habits. Your muscles get stronger during puberty, so you may find that you can do more active things, and get better at them.

HEALTHY HINT

Every day

If you can, try to do one hour of exercise every day to stay fit and healthy.

WHAT TO CHOOSE?

The best kinds of exercise give you stamina and strength and help to keep your body flexible. There are lots of different types of exercise to choose from. Everyone can find an activity they enjoy!

Team sports, such as football, netball and hockey, are great ways to make friends. If you prefer to exercise alone, you could try swimming, skating or cycling. You are more likely to stick at a sport if it is something you find fun to do.

HAPPY TIMES

Experts have found that the brain releases special chemicals, called endorphins, when you are active. Endorphins are hormones that make you feel happy.

Exercise can help you to relax, which in turn helps you to cope better with anything that might be worrying you, including any concerns you may have about puberty. These good feelings can carry on even after you have finished exercising for the day.

HEALTHY HINT

Time to sleep

During puberty, your body works extra hard. Try to get at least ten or eleven hours' sleep a night. This will give your body time to rest and recover.

PUBERTY DIARY

You might find it helpful to keep a puberty diary to record some of the changes taking place during this exciting time in your life.

Keep a record

Use a notebook to record what you notice each day. This could be simple measurements, such as height, or hip or foot size. Note down how your skin and hair are changing, too.

Remember to record how you are feeling. Your emotions are such an important part of the puberty process. Writing down how you feel can be really helpful, both in the moment and looking back at what you have written, later. You could decorate your diary, too.

You only go through puberty once, so having a special record of this time is a great idea. You never know, your puberty diary could be helpful for a younger family member going through puberty in years to come – or even your own children, should you choose to become a parent as an adult!

QUIZ

~~~~~~~~~~

Take the puberty challenge!

1. **Every time your aunt sees you, she tells you how tall you are getting. Do you:**
   a) Ignore her? You hate people talking about the way you look.
   b) Politely ask her to stop commenting on your body?
   c) Tell her to leave you alone because it's none of her business?

2. **You have started your period and you have P.E. at school today. Do you:**
   a) Tell your parent or carer that you feel too ill to go to school?
   b) Make sure you have sanitary products with you that will help you to feel comfortable doing P.E. lessons?
   c) Go to school, but 'forget' to take your kit, so you do not have to take part?

3. **Your friend has started to grow a moustache. Do you:**
   a) Feel jealous, because you have not got any facial hair yet?
   b) Don't say anything? it's a normal part of puberty and you'll catch up, in time.
   c) Laugh at him and tell him he looks silly?

4. **You have a few spots on your face. Do you:**
   a) Feel self-conscious and embarrassed?
   b) Wash your face regularly, and ask your parents to buy some special cream if they get worse?
   c) Squeeze them to try to get rid of them?

5. **You want to go to the park on your own, but your parent or carer says that you are not allowed. Do you:**
   a) Pretend you are doing some homework with a friend, and sneak out to the park without telling anyone?
   b) Tell them that, now you are getting older, you want to be more independent – but listen to them, and work something out that everyone is happy with?
   c) Run up to your room, shouting at them that you hate them?

Answers on page 31

Glossary

acne A skin condition of red pimples that may contain pus, usually found on the face, upper chest and back

carbohydrate A substance found in some foods that gives the body energy

eating disorders Any of several mental health conditions, such as anorexia and bulimia, which negatively affect people's thoughts and behaviours towards food, weight and body shape. Eating disorders are dangerous

egg Female sex cell, from which a baby can develop if an egg is fertilised with a male sex cell – sperm

endorphins Chemicals (hormones) made in the brain that can make you feel happy and relaxed

fertilised When a male sperm joins with a female egg

genitals The private parts of the body between the legs; the penis and testicles, and vulva and vagina

glands Organs in the body that produce chemicals, such as hormones

hormones Chemicals that tell the body to do certain things, for example to grow or sleep

menstrual cup A reusable, small silicone cup that collects blood during a period

nipple A small, raised structure on the front of the breast or chest. In females, the nipple contains glands and ducts (tubes) through which breast milk is delivered to a baby

nutrients Substances found in foods that are good for your health

ovary Female reproductive organs that release eggs

penis Male reproductive organ that ejects sperm

pores Very small holes in the skin that allow sweat to escape the body

protein A substance found in some foods that helps the body to grow and repair itself

puberty the process of body changes from being a child to an adult who can (usually) have children of their own

sanitary pad A disposable piece of soft, absorbent material worn during a period to absorb blood

sebum An oily substance that keeps the skin and hair waterproof and stretchy

self-conscious Feeling very aware of your body, appearance or actions, often in a negative way

silicone A human-made material that is smooth and flexible and able to be shaped for all sorts of uses

social media Websites and apps where users can view, create and share content

sperm Male sex cell. If a sperm joins with a female egg cell, a baby can develop

tampon A small, disposable cylinder of (usually) cotton that fits inside a vagina to absorb blood during a period

testicles Two round body parts that hang below the penis. The testicles produce sperm

uterus The part of a woman's body where a baby develops before birth. It is also called the womb

vitamins Substances found in many foods, which are good for your health

Quiz answers

Mostly bs:
Well done! You have learned a lot about puberty. You know ways to look after yourself, and to respect and get on with others during what can be a roller coaster time. You can find out even more by reading books and websites on puberty topics, if you want to. Always ask permission from a trusted adult before you access the internet.

Mostly as or cs:
You have some way to go before you pass the puberty challenge. Remember, there is a lot going on during puberty. It will help you to find out as much as you can, so you know how to handle it.

Keep reading this book; it will give you lots of information and ideas! And talk to your parents or caregivers – they will be able to give you helpful advice and support, too.

Index